United States Nuclear Regulatory Commission

Protecting People and the Environment

NUREG-1379, Rev. 2

I0493896

NRC Editorial
Style Guide

Office of Administration

AVAILABILITY OF REFERENCE MATERIALS
IN NRC PUBLICATIONS

United States Nuclear Regulatory Commission

Protecting People and the Environment

NUREG-1379, Rev. 2

NRC Editorial Style Guide

Manuscript Completed: May 2009
Date Published: May 2009

Prepared by
C. Hsu

Office of Administration

Abstract

The "NRC Editorial Style Guide" provides writing and style guidance to all U.S. Nuclear Regulatory Commission (NRC) staff members. It addresses the questions and issues most frequently fielded by NRC editors. This revision of the document emphasizes the importance of plain language and includes detailed guidance on using technology to write and edit more effectively. The goals of the "NRC Editorial Style Guide" are readability and consistency for all NRC publications.

Contents

Preface

Editorial style refers to the choices that writers and editors make to eliminate inconsistencies in their documents. These choices are sometimes difficult because language evolves. In updating this style guide, words related to new technologies posed a special challenge. Is it Email, E-mail, e-mail, or email? The U.S. Government Printing Office standard is e-mail, but this guidance clashes with a trend towards closing hyphenated words. The example shows that the authorities who compile style guides do not always agree. Moreover, each discipline has specific standards.

The Office of Administration compiled the revised "NRC Editorial Style Guide" to promote the goals of readability and consistency for all NRC publications. To address frequent staff questions, we have introduced several new sections. Technology and Writing covers key word processing tools for writing, editing, and clearing hidden data from final documents. We have also provided a Helpful Web Reference Tools section to point writers towards resources like a glossary, a list of abbreviations, and NRC Web pages with document templates and more specific guidance. The URLs provided were functional at the time of printing. For the most part, this style guide derives its solutions from the *U.S. Government Printing Office Style Manual*, with a few exceptions that we have noted. We also turned to *The Chicago Manual of Style*, and the *MLA Handbook for Writers of Research Papers* for guidance.

Follow this style guide for most documents, but remember that the NRC is an interdisciplinary agency producing everything from legal rulings to public service announcements. The Office of the General Counsel has its own style guide. NRC press releases are written for the media, and most media outlets follow *The Associated Press Stylebook*. In addition, conferences and scholarly journals often have their own in-house styles that NRC staff members must follow.

As an author, identify your intended audience and determine if that audience has a separate style standard. Most importantly, strive to communicate clearly and comprehensibly with your audience. Even if you are writing for a technical audience, the NRC serves the public and the public has a right to read and understand the agency's documents. Please see tips in the new Plain Language section in this guide to make your writing more clear, direct, and powerful.

For user convenience, we have tried to make this guide a compact, easy-to-use reference. Of course, a guide of this size cannot substitute for the scope of information in the *GPO Style Manual*, which sets the style rules for the Federal Government. Please refer to the *GPO Style Manual* for style information not covered by this guide.

Acknowledgment

We wish to acknowledge the contributions of the NRC staff members who participated in Revision 2 of NUREG–1379, "NRC Editorial Style Guide." Michael Lesar's leadership and management skills made this style guide possible. Malcolm Patterson gave valuable input on the needs of the agency's technical staff. Georgette Price and Catherine Jaegers from the Office of the Executive Director for Operations provided their valuable time and institutional knowledge of traditional NRC practices. Gary Lauffer provided oversight as well as knowledge of U.S. Government Printing Office practices. Keith Azariah-Kribbs contributed with knowledge of major style guides and years of technical editing experience. Cindy Bladey answered difficult style questions. Linda Stevenson provided a rigorous manuscript review. We also acknowledge the NUREG-1379, Revision 1, working group members, listed below. Finally, we thank staff from the Office of Administration who published the "NRC Editorial Style Guide" in October 1989: J.F. Beeson, B.A. Calure, D. Gable, M.T. Lesar, M.F. Mejac, and R.F. Sanders.

NUREG-1379, Revision 1 Working Group	
ADM	Caroline Hsu
ADM	Helen Chang
EDO	Ann Thomas
EDO	Mindy Landau
FSME	Catherine Poland
FSME	Patricia Tressler
HR	James Morris
NRR	Malcolm Patterson
OIS	Paula Garrity
OPA	Ivonne Couret
SECY	Sam Walker

1   Abbreviations and Symbols

Use abbreviations and symbols in technical text, reports, tables, brackets, footnotes, and references. Generally, avoid abbreviations and symbols in the conventional text (e.g., correspondence). Consult NUREG–0544, "NRC Collection of Abbreviations," available at http://www.nrc.gov/reading-rm/doc-collections/nuregs/staff/sr0544/r4/ in the NRC Electronic Reading Room.

Chapters 9 and 10 of the *GPO Style Manual* (http://www.gpoaccess.gov/stylemanual) present a comprehensive list of abbreviations, signs, and symbols and guidance for their use. Follow the guidance in these chapters and the rest of this section unless publication requirements for a particular document dictate otherwise. For example, to publish an article in a journal, you may need to follow the publisher's in-house style. The NRC uses the metric system. If in certain cases you must use nonmetric units, please state the values in metric units first, followed by nonmetric units in parentheses.

1. Use the same form of an abbreviation for both the **singular and plural** forms **of a unit of measure.**

    - 1 m
    - 3 m
    - 1 kg
    - 5 kg

2. Omit internal and terminal **punctuation** unless its omission would cause confusion.

    - 1 in. (Period avoids confusion with the word "in".)
    - 5 cm

3. Use abbreviations for **units of measure** only if they are used with numbers.

    - 200 r/min (*but*—The test would determine the number of revolutions per minute.)

4. State the term from which an **abbreviation not commonly known** is formed, followed by its abbreviation in parentheses. Subsequently, use the abbreviation instead of the term. Redefine abbreviations in every new chapter or major section.

    - iodine–131 metaiodobenzylquanidine (I–131 MIBG)
    - 60 indicated horsepower (ihp)
    - electromotive force (emf or EMF)

5. Use periods in abbreviations for **foreign phrases.** (Et is Latin for "and"; therefore, it is not an abbreviation.)

   - et al. (for et alii, meaning "and others")
   - e.g. (exempli gratia, meaning "for example")
   - i.e. (id est, meaning "that is")
   - et seq. (for et sequentes, meaning "and the following")

6. The abbreviations **e.g.** and **i.e.** (followed by a comma) should be used **only inside parentheses**; otherwise, write out the English equivalents. Do not italicize e.g. and i.e.

   - Today we received specific instructions for preparing the report (**i.e.,** its due date, contents, and format).
   - Today we received specific instructions for preparing the report, **that is**, its due date, contents, and format.

7. Use the **U.S. Postal Service two-letter** State and Province abbreviations in any address or capitalized geographic term.

   - Prince George's County, MD
   - Atlanta, GA
   - *but*—deciduous forests of Wisconsin; black sand beaches in Hawaii

2   Acronyms and Initialisms

Although acronyms and initialisms are types of abbreviations, the guidance for their use differs from that for other types of abbreviations. An acronym is a pronounceable term formed from the initial letters of a compound expression (e.g., LOCA for loss-of-coolant accident). An initialism is a nonpronounceable term formed from the initial letters of a compound expression; the initial letters are pronounced as separate letters (e.g., NRC for the U.S. Nuclear Regulatory Commission).

Most acronyms are capitalized except for certain well-known acronyms that are lower case by convention. For a list of commonly used NRC acronyms, see NUREG–0544, "NRC Collection of Abbreviations," at http://www.nrc.gov/reading-rm/doc-collections/nuregs/staff/sr0544/r4/ in the NRC Electronic Reading Room.

.

1. **Use an acronym or initialism** if the term it represents will be **used more than several times** in a document. The first time a term is used, state the words from which the acronym is formed, followed by the acronym in parentheses. As a general rule, you can subsequently use the acronym instead of the term. In a lengthy document, **restate the term followed by its acronym at the beginning of each chapter** or major section. However, keep in mind that **excess acronyms and abbreviations can make an otherwise well-written document difficult to understand**. When writing for a lay audience, or even a technical audience that does not share your specialty, use acronyms carefully. You may sometimes need to redefine an acronym, offer a brief explanation of the acronym, or use a synonym instead. Always **avoid stringing several acronyms together in a single sentence**. Do not include acronyms in headings and titles.

   - The Resource Advisory Council (**RAC**) has reviewed this plan. The **RAC** advised the staff to revise three sections.
   - The Resource Advisory Council has reviewed this plan. The **council** advised the staff to revise three sections.

2. Although an acronym or initialism is capitalized, **do not capitalize** the term it represents **unless** the term would **ordinarily** be **capitalized**.

   - Office of the General Counsel (OGC)
   - technical specification (TS)
   - crack opening displacement (COD)

3. **To form** the **plural** for most acronyms and initialisms, add a lower case *s* without an apostrophe.

   - 12 NPPs
   - five RGs
   - 2.0 FTEs

4. **To form** the **possessive** of an acronym or initialism, use an apostrophe plus s, just as you would for a normal word.

- EDO's report
- IAEA's May conference
- RES's funding (*also*—funding from RES; RES funding)
- O_3's chemical instability (*also*—ozone's chemical instability; the chemical instability of O_3)

5. **To decide** whether *a or an* should precede an acronym or initialism, pronounce the first syllable of the acronym. "A" should precede a consonant sound; "an" should precede a vowel sound.

- an ACRS meeting ("ay" is a vowel sound)
- an AEC report ("ay" is a vowel sound)
- a FEMA decision ("fee" is a consonant sound)
- an NRC office ("en" is a vowel sound)
- a LOCA occurred ("low" is a consonant sound)

3   Capitalization

Capitalize the first word of a sentence, the pronoun I, and proper names. When in doubt, do not capitalize. For example, if a term could be either a proper name or a description (e.g., the department of engineering or nuclear power plant), using the lower case is correct.

1. Capitalize **titles of persons if they precede a personal name**. Use lowercase if the title follows the name except for titles of great eminence, including heads of state, assistant heads of state, heads of governmental units, and royal rulers. Do not capitalize general references to titles. Likewise, **capitalize specific names of organizations** but not general references to them.

    *   President George Washington (*and also*—the President)
    *   King Bhumibol Adulyadej (*and also*—the King of Thailand; His Royal Highness)
    *   Governors Richardson and Lingle (*and also*—The Governors will meet later.)
    *   U.S. Senator Daniel Inouye (*but*—Daniel Inouye, senator; the senator)
    *   The staff in Region IV (*but*—the regional office)
    *   Judge Mia Jones (*but*—Mia Jones, the judge)

    For NRC titles, follow the above guidance, but you may capitalize titles of branch chief and above. This NRC practice differs significantly from *GPO Style Manual* Rule 3.35. Do not capitalize general references to most titles.

    *   NRC Senior Health Physicist Ana Lee (*but*—two NRC health physicists)
    *   Deputy Director Ryan Yamada (*and also*—Ryan Yamada, Deputy Director)
    *   NRC Director of Human Resources, David Calvo (*and also*—David Calvo, Director of the Office of Human Resources)

2. Capitalize **Commission or Commissioners** when referring to the collegial head of the NRC as a group and **Chairman or Commissioner** when referring to the Chairman or a member of the Commission individually. This NRC practice differs from *GPO Style Manual* Rule 3.35.

    *   The Chairman requests a response by Friday and expects the Commission to discuss the issue Monday morning.

3. Capitalize a well-known short form of a specific proper name.

    - the Congress (U.S. Congress, *but*—congressional action, congressional staff)
    - the President (of any country)
    - the District (District of Columbia)
    - the Capitol (in the District)
    - the Agency (for U.S. Central Intelligence Agency, U.S. Environmental Protection Agency, *but*— the agency for U.S. Nuclear Regulatory Commission and all other organizations that do not have "Agency" in their official title)
    - the Office (U.S. Government Accountability Office, but—the office for all other offices, including NRC internal offices)

4. Capitalize the following governmental organizations.

    - Federal
    - Federal Government
    - State (*but*—do not capitalize "local")
    - Nation as a synonym for United States as in "10 percent of the Nation's energy supply" (*but*—"nation" for general references such as "a nation devoted to prosperity")
    - Federal, State, and local responders
    - Tribal officials (Unlike GPO, the NRC capitalizes "Tribal" to signify a sovereign entity.)

5. Capitalize **a common noun followed by a letter or number** that refers to a specific publication, class, figure, or table, except for page or paragraph. (This NRC practice differs from *GPO Style Manual* Rule 3.9.)

    - Category I
    - IAEA Category III sources
    - Appendix K
    - Chapter III
    - Class 1E
    - Table 4
    - Figure 8
    - Section 7
    - *but*—paragraph 4 on page 82

6. Capitalize the first letter of the symbol for an isotope, chemical element, or compound. Do not capitalize the spelled-out name for the symbol.

    - Na   sodium
    - NaCl   sodium chloride
    - Co–60   cobalt–60

7. Capitalize a **trade name**.

- Xerox
- Halogen
- Plexiglas
- Vu–Graph
- Zircaloy (*but*—zirconium)

8. Capitalize the names of all **computer codes**, and capitalize the names of computer **languages and software** consistent with their trade names. If the code is an acronym, spell out the full name the first time it appears in a document.

- Code: MESORAD, CPLUME
- Language: Fortran, Oracle
- Software: dBase, Lotus 1-2-3

9. Capitalize a descriptive term that denotes a **geographic region or feature** used as a proper name.

- the Midwest
- the Continental Divide
- the Deep South
- Western Europe
- Gulf States (*but*—gulf coast)

10. Do not capitalize a term used as a **geographic direction or a position** that is not a proper name.

- northerly, northern
- north, south, east, west
- eastern seaboard
- west coast, east coast

11. Capitalize **months**, but not **seasons**.

- May, October
- spring, autumn
- spring 2007 (*not*—Spring 2007)

12. Capitalize the first word and all important words in **titles of publications and legislation**.

- Secrets of the Past: Nuclear Energy Applications in Art
- The Paperwork Reduction Act of 1982

13. Do not capitalize **articles, prepositions (except** for "**To**" as part **of an infinitive**), and **conjunctions** unless they are the first word of a title or important to its meaning.

   - *How To Write, Speak, and Think More Effectively*
   - *Creating Energy Sources for the Future*
   - *A Guidebook to Nuclear Reactors*
   - "Handbook of Nuclear Safeguards Measurements Methods"

14. Capitalize **hyphenated words** by the same rules you would use if the words were not hyphenated.

   - ex-President Clinton
   - Commissioner-elect Jones
   - high-level waste
   - "A Study of Proposed High-Level Waste Repositories"

15. Do not capitalize the names of systems at nuclear facilities.

   - residual heat removal system
   - emergency core cooling system

16. **Do not capitalize document types**, unless followed by a number or letter referring to a specific document.

   - regulatory guide (*but*—Regulatory Guide 1.18)
   - management directive (*but*—Management Directive 4.1, "Accounting Policies and Practices")

17. **Do not capitalize** a common noun followed by a letter or number identifying a **component of a nuclear power plant**.

   - train A
   - valve PRV–22

18. **Capitalize basic alphabet keys**, all the **named keys** (e.g., Ctrl, Shift, Alt) and **menu items** (e.g., File, Save, Print).

   - To run a spell check, go to Tools, then choose Spelling and Grammar.
   - A keyboard shortcut for copying text is Ctrl+C.

19. Capitalize these **technology-related** words. For information on how to compound technology-related words, see Section 4.

   - Internet
   - Web page
   - Web site
   - Web

4   Compound Terms and Unit Modifiers

A **compound term** can appear in one of the three following forms: (1) open (e.g., nuclear power plant, mill tailings site), (2) closed (e.g., rulemaking, runoff), and (3) hyphenated (e.g., full-scale simulation, fitness-for-duty program). Although there are rules to compounding terms, there are also many exceptions. Modern English is moving towards a general trend of closing compounds, unless doing so causes confusion. For example, the modifier on-line or on line, is now online.

Chapter 6 of the *GPO Style Manual* presents general rules for compounding words. To look up a specific word, go to Chapter 7 of the *GPO Style Manual*, "Compounding Examples," where you will find 6,000 compound words in their correct open, closed, or hyphenated form. To find the *GPO Style Manual* online, go to http://www.gpoaccess.gov/stylemanual.

Compound terms that modify nouns are called **unit modifiers**. Those that precede nouns are typically hyphenated. Those that follow the nouns they modify are typically not hyphenated.

- an NRC-sponsored study (*but*—a study sponsored by the NRC)
- an industry-sponsored study (*but*—a study sponsored by industry)
- handled on a case-by-case basis (*but*—handled case by case)

Use hyphens carefully with unit modifiers because their placement can cause misreadings. For example, depending on the usage of the words, each of the following phrases or sentences is correct.

- a biological-waste management system
- a biological waste-management system

The following table provides detailed guidance on the use of hyphens with unit modifiers.

1. Hyphenate **unit modifiers** used as adjectives and adverbs that precede a noun. Some **types and examples** follow:

Type	Example
Modifier plus present participle	far-reaching effects hard-working staff fear-producing accident thought-provoking analysis

Type	Example
Modifier plus past participle	safety-related activities coal-fired plants performance-based incentives well-defined plan much-acclaimed study long-lived isotope
Multiple-word modifier	not-in-my-backyard attitude easy-to-read document loss-of-coolant accident boiling-water reactor pressurized-water reactor nozzle-to-pipe weld
Suspended modifier (i.e., use of a unit common to a series of unit modifiers)	industry- and agency-sponsored studies long- and short-term goals 8-, 9-, and 10-hour days
Unit modifiers in which one unit is capitalized	PDR-available documents NRC-prepared report
Unit modifiers that include numbers (see also Section 5)	18-inch-diameter pipe six-person team two-shift operation
A prefix before a proper noun, capitalized abbreviation, or number	non-Federal anti-American pre-Jurassic pre-NRC regulatory agency post-2001 security measures
A few adjective-noun unit modifiers	high-level waste low-pressure injection light-water reactor nonlight-water reactor nonsafety-related component

2. Do not hyphenate (leave open) unit modifiers ending in *ly*, as well as certain other modifiers. Some **types and examples** follow:

Type	Example
Modifier ending in *ly*	poorly managed plant effectively managed branch
Unit modifiers both elements of which are capitalized	an Appendix A criterion 10 CFR Part 50 requirements

Type	Example
Foreign phrases used as unit modifiers	ad hoc meeting ad hominem argument in vitro tissue culture in situ hybridization
Three-word unit modifiers whose first two words are adverbs	very well defined procedures unusually well maintained design basis
Noun formed from a one-syllable verb and adverb	makeup heatup cooldown startup shutdown followup

3. Close up most prefixes, but be aware of many exceptions. Use a hyphen to avoid mispronunciation or confusion. Some **types and examples** follow:

Type	Example
Prefixes	multiplant prelicensing semiannual biweekly nondistinctive nonsafety overpressure
Use a hyphen before a proper noun, capitalized abbreviation, or number.	pre-Columbian artifacts pre-2001 security measures post-U.S.S.R.
Use a hyphen with prefixes to avoid doubling a vowel or tripling a consonant, except after *co, de, pre*, and *re*. Also use a hyphen to avoid mispronunciation or ambiguity.	anti-inflammatory gull-like coowner deenergize re-creation vs. recreation multi-ply vs. multiply pre-position vs. preposition co-op un-ionized vs. unionized

4. Close most cases of *wide* and *making* used as a suffix.

- agencywide (*but*—NRC-wide because it is a proper noun)
- industrywide
- worldwide
- decisionmaking
- rulemaking
- policymaking
- Governmentwide

5. Usually, do not hyphenate **chemical and physical terms** except for chemical formulas.

- boric acid solution
- carbon monoxide gas
- equivalent uranium content
- hydrogen ion activity
- ground water
- Cr-Ni-Mo

6. Follow these conventions for **technology-related words**. Be aware that conventions may change. For guidance on capitalizing technology-related words, see Section 3 of this guide, "Capitalization."

- Web page
- Web site
- Web-related issues (adjective form is still hyphenated)
- Web broadcasting
- Webcasting
- e-mail
- online
- database

5   Numbers

1. Spell out the numbers **one through nine**.

    - Four reactor licensees reported seven events.
    - The crew replaced four pipes, six valves, and nine gears in record time.
    - *not*—The crew replaced four (4) pipes, six (6) valves, and nine (9) gears in record time.

2. Use numerals for a single number of 10 or more.

    - During August 1985, 21 plants submitted the required reports, but of these reports, two were late.
    - About 30 engineers attended the meeting.

3. Spell out a **number that begins a sentence**, and spell out related numbers at the beginning of a sentence separated by no more than three words. Alternatively, revise the sentence to avoid beginning the sentence with a number. In these next examples, assume that the author has already defined Title 10 of the *Code of Federal Regulations* (10 CFR) Part 20, "Standards for Protection against Radiation."

    - As required by 10 CFR Part 20, radiation protection plans for certain facilities. (*but*—do not begin a sentence with "10 CFR Part 20 requires...")
    - Twenty families face possible evacuation.
    - Seventy or, perhaps, eighty square miles were affected by the aftershock. *or*—The aftershock affected approximately 70 or 80 square miles. (*but*—Eight acres of the 12 square miles affected by the aftershock were within the site boundary.)

4. When two or more **related numbers** appear in a sentence **and one** of them **is 10 or more**, use a numeral for each number.

    - The inspectors found fractured pipes in four plants: 6 in Fermi, 10 in Watts Bar, 2 in Monticello, and 4 in Susquehanna.
    - The NRC received comments from 13 utilities, 18 public interest groups, 3 unions, 6 utility organizations, and 2 interested persons.

5. Use numerals to express a unit of measurement, such as **time or money**. This usage does not affect other numerical expressions in a sentence.

    - 2 meters
    - 15 years
    - 3.5 percent
    - 6-cm-diameter pipe

- Three NRC travelers left the office at 3:30 p.m. for a 4-day trip to join an eight-person advisory committee on radiation safety. Each traveler had to travel approximately 50 miles per day and each had a $600 travel advance. On this trip, a business day lasted about 10½ hours—beginning at 8 a.m. and ending about 6:30 p.m.—with a 1-hour break for lunch. Of the 8 members of the committee, 4 supervised a staff of 11 or more health physicists.

6. In tables and illustrations, use **numerals with abbreviations, symbols, decimals, and with quantities that mix whole numbers and fractions**. In text, do not use symbols. Write out the abbreviation on first appearance in text and thereafter use the abbreviation.

Tables and Illustrations	Text
3 km	3 kilometers
12 cm	12 centimeters
3′–6′ *	3 feet to 6 feet
12″ **	12 inches
8″ x 12″	8 inches by 12 inches**
1,500 psi	1,500 pounds per square inch
33%	33 percent
3 kg	3 kilograms
1.8 g	1.8 grams
1½ r/min	1½ revolutions per minute
100 °C	100 degrees Celsius***

*	Note that the symbol ′ is **prime** (Unicode character 2032), not the apostrophe.
**	Note that the symbol ″ is **double prime** (Unicode character 2033), not the closing double quotation mark.
***	Note that the symbol ° is the **degree** (Unicode character 00B0), not the ring above diacritic.

7. Spell out **indefinite numerical expressions**—"about," "nearly," "around," and "approximately" are not considered indefinite (see the example in Item 5 of this section).

- the early seventies (*but*—nearly the 1970s)
- midforties (*but*—around 1945)
- seemingly a hundred and one reasons (*but*—he cited about 10 reasons)

8. An ordinal number expresses degree or sequence. Apply the general rules for numbers in this section to **ordinal numbers**.

- The accident rate for the fourth quarter suggested that the training had been effective.
- Surprisingly, the 19th and 20th years of plant operation produced the highest income.
- The *Federal Register* notice was published on the 27th of March (*not*—March 27th).

9. When **two numbers** appear **in sequence**, use a numeral for one and spell out the other.

- The inspector examined twelve 12-inch-diameter pipes.
- The fold-out page consisted of eleven 2-inch columns of numerals.

10. Use numerals in all mathematical expressions.

- multiplied by 4
- a factor of 9

11. Spell out a fraction standing alone, a fraction followed by "of a" or "of an," and a fraction approximation.

- three-fourths of a mile
- seven-eighths of an inch
- The water on three-fourths of the site was contaminated.
- The team leader has reviewed nearly three-fourths of the inspection report.

12. For a spelled-out fraction, use a **hyphen** between the **numerator and denominator**. However, omit the hyphen between the numerator and denominator when a hyphen already appears in either or both.

- three-fifths
- six thirty-fifths
- twenty-one thirty-fifths
- twenty-three thirtieths
- two one-thousandths

13. Use numerals for a fraction in a unit modifier.

- ½-inch width
- ¾-mile radius

14. Use numerals when combining **whole numbers and fractions**.

- 2½ times as large
- 1½ inches wide

15. Use numerals for all **decimals**.

- 0.5 mL
- 1.8 meters

16. Use a **decimal point** only for **monetary amounts** that include cents, but omit for even dollar amounts.

- $150.10
- $150

17. For quantities of less than one, use a **zero before the decimal point** unless the quantity could never reach one, such as the caliber of a gun. Use zero **after a decimal point** if the zero is followed by other nonzero numbers or if it is a significant number; that is, the zero is holding a place. Do not use a decimal place followed by a zero after a whole number (unless required for precision).

- 0.5 part (*but*—.44 caliber)
- 5.04
- 5.003 (*but*—5)
- 45 (*not*—45.0)

18. Use **subscripts** to indicate the base for a number system and the number of atoms of an element in a molecule.

- 25_8
- H_2

19. Use **superscripts** for exponents and to indicate the mass number of an isotope (see also Section 3, Item 6, of this guide).

- 2^5
- ^{235}U (*but*—uranium-235)

20. Generally, do not use the calculator or computer printout abbreviation for an **exponent** in text or tables (**i.e., use the superscript**). Use of computer printout abbreviations is acceptable for computer codes. If you must use the computer printout abbreviation, please apply it throughout the document. Do not switch between the computer printout abbreviation and superscripts.

- 8^3 (*not*— 8**3)
- 8×10^3 (*not*— 8E3)
- 8×10^{-3} (*not*—8E-3)

21. Use **commas every three places** starting at the right **to separate numbers** over three digits.

- There were 5,000 comments on the proposed rule.
- There were 50,000 responses to the questionnaire.
- There were 3,000 responses to the first questionnaire and 15,000 to the second.
-

Table
5,000
50,000
1,000,000

22. To aid comprehension, spell out **million or billion** (but not always in tables and figures).

- $12 million
- $1.25 billion
- $5 million to $10 million
- 4.2 billion years
- population of 2.8 million

23. Use the **percent symbol** in tables, graphs, and figures. Otherwise, spell out percent in the text.

- The office had expended 80 percent of its budget by midyear.
- Of the 103 reactors, 75 percent had replaced at least one steam generator after 10 years.

24. When writing about a **range of numbers**, give the **full digits**. Use an en dash to separate numbers in a range.

- ages 25–28 (*not*—ages 25–8)
- pages 1260–1268 (*not*—pages 1260–68, and *not*— pages 1260–8)
- days 104–107 (*not*—days 104–7, and *not*— days 104–07)
- from years 2006–2007 (*not*—from years 2006–07, and *not*— from years 2006–7)
- January 2–9, 11, and 13–15 and March 5–9, 26, and 30, 2007

6   Punctuation

1. Use a **comma after each member of a series** of three or more words, letters, numerals, phrases, or clauses.

    - footnotes, references, and bibliographies
    - between A, B, or C
    - neither in 1999, 2001, nor 2007
    - in the morning, in the afternoon, but not in the evening
    - The section leader wrote the proposal, the branch chief reviewed it, and the division director signed it.

    Change the usual commas in a series to semicolons if multiple commas occur within the elements of the series.

    - After assessing a situation, the NRC may order a licensee to continue, curtail, or expand operations; ensure compliance with security and safeguards programs; and maintain associated notes, documents, and records of these emergency actions.

2. Use a **comma after an introductory phrase** of five or more words; use of a comma after a shorter introductory phrase is optional.

    - After realizing that they could not adhere to the day's agenda, they rescheduled three topics for discussion.
    - Next Tuesday the proposals should arrive. (*or*—Next Tuesday, the proposals should arrive.)

3. Use a **comma after an introductory adverbial clause**.

    - After struggling with the problem, he decided to consult experts.
    - Before moving onto my next location, I would like to meet with you.

4. Place a **comma before a direct quotation** of a few words.

    - He said, "The tube has been welded."
    - "But," the welder said, "it is only a temporary measure."

5. Use a **comma before and after an explanatory equivalent** of another word or phrase.

    - Nishi Mackin, president of TRCO, met with the Chairman.
    - The American National Standards Institute has published a more recent standard, "Scientific and Technical Reports: Organization, Preparation, and Production," ANSI Z39.18–1987.

6. Use a comma before and after the year when written in the order of month, day, and year.

- The March 27, 1988, memorandum responded to the questions. (*or revise to read*—The memorandum, dated March 27, 1988, responded to the questions.
- On March 27, 1988, he attended the conference.

Do not use a comma in a two-element date or a three-element date written in the order of day, month, and year.

- March 27
- March 1988 (*not*—March, 1988)
- 27 March 1988 (military usage)

7. Use a **comma before and after the State or country** when citing the city and State or city and country in the text.

- The meeting is in Chicago, Illinois, on April 4, 2008, at 3 p.m.
- A conference in Paris, France, would cost more than one in Paris, Texas.

Do not use a comma **between the State and** the **ZIP Code** in an address.

- Bethesda, MD 20014

8. Use a comma **before and after *Jr., Sr., Inc.,* and *Ltd.*** (Some individuals or companies choose to omit the comma, and the NRC can follow their style at its discretion.)

- The staff will meet with Chem-Nuclear Systems, Inc., on September 5, 1988.
- We know that Time Inc. produced the book.
- Ganesh Gritz, Jr., chaired the meeting.

9. Omit the comma before and after II, III, IV, etc. in a name and between the State and the zip code.

- Henry VIII
- John Francis Kipp III
- Rockville, MD 20850

10. Use a **comma between qualifying words** if the word *and* could replace the comma.

- An old, degraded generator tube (*but*—an old generator tube)

11. Use a **comma before the conjunction** in a compound sentence.

- The agency received the application in October 2007, but staff did not approve the terms until January 2009.
- I plan to arrive at the site on Tuesday, and I am scheduled to begin the inspection on Wednesday.

12. Use a **comma to set off nonrestrictive words, phrases, or clauses** that could be omitted without changing the meaning of the sentence.

   - The final rule, which was published on December 13, became effective on January 12.
   - Members of the public, including those from the eastern districts, attended the hearing.

13. Use a **comma after "however"** at the beginning of a sentence, unless "however" is used conditionally.

   - However, he continued to smoke.
   - However, I would still like to discuss the topic.
   - However much he spoke, he never seemed to get to the point.

14. Use a **semicolon** to separate closely related or contrasting statements. Use one space after a semicolon.

   - He agrees; I do not.

15. Use a **semicolon before an adverb** (e.g., then, however, thus, hence, indeed, accordingly, besides, and therefore) **joining** independent clauses.

   - I attended the meeting; however, I had to leave before it ended.
   - He had doubts about the public hearing; indeed, it was pure chaos.

16. Use a period at the end of a sentence. Add two spaces between sentences in all NRC correspondence. For any other document, follow the style guidance provided by the office for which it is prepared. Use a period after items in a list if each item is a complete sentence.

   - She must return on Friday. The following is her itinerary:
     1. Monday, she flies to New York.
     2. Tuesday, she chairs the conference.
     3. Wednesday, she conducts a hearing.
     4. Thursday, she addresses the council.

17. Use a **colon after** a complete **clause to introduce a list**, whether or not the list is within a sentence. For all NRC correspondence, when using a colon within a sentence, follow it with two spaces. For any other document, apply the style guidance provided by the office for which it is prepared.

   - Greg Emes was responsible for the following:
     1. basic research
     2. confirmatory research
     3. written results
   - The judges consider three factors: taste, texture, and aesthetic appeal.

18. Use a **colon after a formal salutation**. Please consult Management Directive 3.57, "Correspondence," for more guidance on letters and salutations.

- Dear Mr. Chairman:
- Dear Mrs. Leboulle:
- Dear Senator McCain:

19. Use a **colon before a lengthy quotation**. Indent a quotation of five or more lines on the right and left and omit the quotation marks. The indentation of the material indicates that it is a verbatim quotation.

- The excerpt from Bulletin 4145–220 follows:

  > The purpose of the program is to provide income protection to employees affected by a medical emergency through the voluntary donation of annual leave by other employees.

20. Use an **em dash to mark an abrupt change** in thought. (In Microsoft Word, the keyboard shortcut for em dash is Ctrl+Alt+Minus Sign. Using the toolbar, insert an em dash by clicking on Insert, choosing Symbol, clicking on the Special Characters tab, and finally choosing the em dash.)

- Provide several publications—in addition to your resume—to highlight your qualifications.

21. Use **em dashes to replace commas around an interrupting element** with heavy internal punctuation.

- The group—engineers, managers, and administrators—toured the facility.

22. Use an **en dash to indicate ranges of number, letters, or periods of time**. (In Microsoft Word, the keyboard shortcut is Ctrl+Minus Sign. Using the toolbar, insert an em dash by clicking on Insert, choosing Symbol, clicking on the Special Characters tab, and finally choosing the em dash.)

- For several years (2003–2006), the company has downsized their staff.
- The planned September–December trip never happened.
- From January 21–March 5, 2008, the licensee closed the plant for substantial repairs.

23. Use **parentheses to show explanatory information**. Punctuate a sentence with **parentheses** the same as a sentence without parentheses. Do not precede an opening parenthesis with a comma.

- The inspector visited three of the four regions (I, III, and IV).
- Before leaving the site (late Friday afternoon), he submitted his report.
- The high-level waste regulation (10 CFR Part 60) establishes the requirements for site characterization of a geologic repository. (The U.S. Department of Energy will be the only applicant to use the part.)

24. Use **brackets within parentheses or parentheses within parentheses to nest information**. Also use brackets to indicate words you have inserted into a direct quotation.

- Please review the applicable international standard (Nuclear Power Standard 6.5 [NP-STD-6.5]).
- "Some time last night [the] idea came to me," said Mr. Pai.
- He has not reached the annual whole-body limit for workers (50 millisieverts (mSv)).

25. Place the **period or comma inside a quotation mark**.

- He said, "The project is due today."
- He said, "The project is due today," as he rushed to his office.
- The center is offering three courses today: "MS Outlook," "Reactor Core Concepts," and "Web Design 3."

26. Place a **question mark, colon, or semicolon outside** the closing **quotation mark** (unless it is part of the quoted material).

- We hired the most qualified environmental policy analyst from the "Presidential Management Fellows Program"; we did not realize he lacked motivation.
- He said, "The project is due today"; he rushed to his office.
- Have you seen my copy of NUREG–1379, "NRC Editorial Style Guide"?
- *but*—I recommend you read the article "Nuclear Energy: A Way to A Greener Future?"

27. Place other punctuation marks inside the closing quotation mark only if they are part of the matter quoted.

- Did you mark the package "Fragile"?
- She asked, "Do you have sufficient funds to cover the travel?"

28. Use **quotation marks** to emphasize special technical or industry-specific language, letters, names, words, or titles.

- The flat strips were "pigtailed" to create tubes.
- Faced with plan "A" and plan "B," we chose plan "B."
- She frequently cites Title 10 of the *Code of Federal Regulations* Part 8, "Interpretations."

29. The **slash is ambiguous** and means *"and," "or,"* or both. **Use slashes in tables and graphs** for brevity, **otherwise avoid using slashes whenever possible**, especially if you really mean either *"and"* or *"or."* Avoid writing and/or.

- Notify your supervisor if you will be late or absent. (*not*—Notify your supervisor if you will be late and/or absent. *not*—Notify your supervisor if you will be late/absent.)
- The center serves clients who are elderly or disabled, or both. (*not*—The center serves clients who are elderly/disabled.)

30. Use a **slash to join multiple-word unit modifiers**.

- A joint U.S. Nuclear Regulatory Commission/U.S. Department of Energy initiative (*but*— a joint NRC–DOE initiative)

31. Use a **slash to indicate "per."**

- 60 km/h
- 11 pills/day

32. **Avoid breaking a URL** (uniform resource locator) over two lines. If necessary, the break should be between elements after a colon, slash, or double slash, but before a period. Do not use a hyphen to break a URL. A URL that contains a hyphen should never break at the hyphen.

- The best option is to keep the URL, such as http://www.internal.nrc.gov/, on one line.
- However, if you must break the URL, break along a colon, slash, or double slash http:// www.internal.nrc.gov/ as we have done on this line.
- If you must break the URL near a period, break it before the period http://www.internal .nrc.gov/ as we have done on this line.

33. **Use italics for foreign words and phrases that are not well known** to English speakers. Do not italicize commonly used foreign words, including scholarly Latin. Do not italicize foreign proper nouns.

- *Belgique c'est chic* was his personal motto.
- An earthquake hit the Kashiwazaki-Kariwa Nuclear Power Plant in Niigata Prefecture.
- She did not seem sorry about eating all the foie gras.
- *Schadenfreude* means taking joy in the misfortune of others.
- ibid., et al.

34. Use an apostrophe before an "s" to indicate possession.

- Cameron's policy
- Phyllis's plant (*also*—her plant)
- U.S. Congress's history (*also*—the history of the U.S. Congress)
- Dr. Seuss's hat (*also*—the hat of Dr. Seuss)

35. Use lists to organize text. If the list is set in the body of the text, completely enclose numbers or letters in parentheses. Use numbers for sequential list items or list items you will refer to later. Use bullets for random list items. For short lists, run the list into the text (unless you are writing codified text).

He had three resolutions: (1) to learn French, (2) to pass the French proficiency exam, and (3) to find a job in Brussels, Belgium.

36. For **longer lists**, set the list vertically. You may leave **incomplete sentences** unpunctuated. Always use **parallel sentence construction** for lists.

The working group is making progress in the following primary goals:

- searching for a director with government experience
- reorganizing top management into three primary areas
- bidding out contracts on printing
- redrafting the mission statement for emphasis on public accountability

37. Unless you are writing a legal document or following a specific convention, **avoid punctuating lists as a running sentence with commas or semicolons**. However, if you must use a running sentence list, note the **first word of each list item is not capitalized**.

The working group is making progress in its primary goals of—

- searching for a director with government experience;
- reorganizing top management into three primary areas;
- bidding out contracts on printing; and
- redrafting the mission statement to emphasize public accountability.

38. **If one list item is a full sentence with a period, then all list items should be full sentences with periods.**

The working group is making progress in the following activities:

- It is searching for a director with government experience.
- It is reorganizing top management into three primary areas.
- It is bidding out contracts on printing.
- It is redrafting the mission statement to emphasize public accountability.

7   Plain Language

The Plain Language Initiative was one of several conceived by the National Partnership for Reinventing Government. The goal of the partnership was to create a Government that "works better, costs less, and gets results." For more indepth guidance on using plain language, visit the NRC Plain Language Writing Techniques Web page at http://www.internal.nrc.gov/ADM/DAS/cag/notices/notdocs/writingtech.html.

Remember that simple language is easier to read and is often more accurate. Give the correct level of detail that your audience needs. Do not inundate a layperson reader with technical details. For documents meant for public comment, aim for a grade 9 to 12 reading level. To check reading level in Microsoft Word, use the readability statistics function in Spelling and Grammar. (See Section 12, "Technology and Writing," for instructions.)

1 Principles of Plain Language

- Use reader-oriented writing. Write for your customers, not for other Government employees.
- Use natural expression. When possible, write as you would speak. Write with commonly used words.
- Make your document visually appealing. Present your text in a way that highlights the main points.

2 Tactics for Clearly Communicating Your Message

- State the purpose of your document.
- Explain how you have organized it and how to use it.
- Insert descriptive headings to help the reader find information.
- Summarize complicated topics before you give the details.
- Begin with matters of most interest to your reader.
- Answer general questions first.
- Describe processes step by step.
- Omit information the reader does not need to know.

3 Rules for Writing in Plain Language

1. Use the active voice.

    The active voice tells who is supposed to do what. It is explicit and direct, assigning a clear agent to an action.

    - *Active*—The licensee installed a new steam generator.
    - *Passive*—A new steam generator was installed.
    - *Active*—The staff held a conference call with the BWR Owners Group.

- *Passive*—A conference call was held with the BWR Owners Group.

Use the passive voice only when the object of an action is more important than the agent.

- *Acceptable Passive*—Taking photographs is forbidden in this secure area.
- *Active*—Malcolm A. McDuffy, Chief, Division of Security, forbids the taking of photographs in this secure area.

2. Write in short sentences and short paragraphs.

Express only one idea in each sentence. Short sentences are better for conveying complex information. They break the information up into smaller, easy-to-process units. If your sentence is **more than three lines long**, **try to break it into two or more separate sentences**. The more technical the topic, the shorter the sentences should be to aid in reader comprehension.

Short paragraphs break up material into easily understood segments and are visually more appealing. They also allow you to insert more informative headings in your material.

- *Long:* Associated with amending 10 CFR 50.44, a conference call was held with the BWR Owners Group (BWROG) with the purpose of exploring possible coordination between the staff's activities and related efforts currently being pursued by the BWROG.
- *Short:* The NRC staff and the BWR Owners Group held a conference call to discuss coordinating their activities in amending 10 CFR 50.44.

3. Use as few words as possible. Simple language is often more accurate.

Instead of	Use
a majority of	most
in a timely manner	promptly
a number of	some
no later than	by
at this point in time	now

Instead of	Use a single-word verb
arrive at a solution	solve
be cognizant of	know
held a meeting	met
hold a discussion	discuss
make a determination	determine
perform an analysis	analyze

Instead of	Use a single-word verb
adversely impact	impair, damage, compromise, decrease, increase

4. Use plain words.

Instead of	Use
a and/or b	a or b or both
assist	help
conduct	do, carry out
feasible	workable, doable
finalize	complete, finish
identify	find, note
impact	affect, change
indicate	say, state
methodology	methods
numerous	many
perform	do, study, analyze
preclude	prevent
prior to	before
provide	give, supply
regarding	about, on
request	ask
subsequent to	after
utilize	use

5. **Avoid legalisms**. Even if you are writing a legal document, use simple everyday words. Avoid the following vague legalistic terms:

- aforementioned
- hereby
- herein
- hereinafter
- therein

Instead of	Use
abeyance	postpone action
afford an opportunity	allow, permit
finalize	end, conclude
fullest possible extent	completely
implement	carry out
promulgate	issue
pursuant to	under
terminate	end
verification	verify

6. Avoid confusion by observing the following.

- **Avoid excess abbreviations**. To reduce acronyms, use a key word from the term instead of the acronym. Use key words to avoid multiple abbreviations in one sentence.
  - The Center for Nuclear Waste Regulatory Analyses submitted its initial draft report on time. The center agreed to submit the final report on December 10, 2000.
  - *Not*—RES agreed with NRO that the NRC should streamline the COL process. (*use instead*—Research agreed with NRO that the agency should streamline the combined license process.)
- **Use consistent terms** throughout the document.
  - If you start with Reactor Oversight **Process**, do not switch to reactor oversight **program**.
  - If you begin by calling it an **assessment**, do not switch to **evaluation**, **analysis**, or **study**.

- Unbundle strings of five or more nouns.
  - *Write*— procedures to protect the quality of surface water (*not*— surface water quality protection procedures).
  - *Write*— the threshold value at which the primary coolant must be sampled (*not*— primary coolant sampling requirement threshold value).
- Make sure **pronouns have clear antecedents**. In the following example, "it" could refer to safety margin, measure, or design.
  - *Clear*—Safety margin is a measure of the conservatism employed in a design to ensure that the design will work.
  - *Vague*—Safety margin is a measure of the conservatism employed in a design to ensure that **it** will work.
- **Avoid technical terms that only a specialist would understand** unless you are writing for a purely technical audience. If you must use the technical term, add a few words to explain the meaning.
  - *Technical*—The printing specialist is reviewing the bluelines.
  - *Plain Language*—The printing specialist is reviewing a proof of the publication that is printed in blue ink.
  - *Plain Language with Explication*—The printing specialist is reviewing a proof of the publication known as "the bluelines" because it is printed in blue ink.

7. **Use the pronouns I, we, and you to speak directly to the reader**. Especially in correspondence, this establishes an informal tone. More formal third-person usage is often used in formal writing, such as in technical reports.

   - I am responding to your letter of April 2, 2000, about the use of potassium iodide to reduce the uptake of radioiodine in the event of a nuclear accident.
   - Avoid using the royal "we."

8. Use bullets, lists, tables, and graphics for visual variety. Shorter lines and more white space on a page give the reader's eye a rest from dense blocks of text. A list is easier to read than running text. Use numerical lists if the items have an order of importance or if your introduction to the list identifies a specific number of items, as in this example.

   - The management of mining and milling residues, such as tailings and waste rock, is also outside the scope of this publication. However, the publication covers the decommissioning of facilities and equipment for surface industrial extraction associated with mining and milling. Fuel cycle facilities pose the following four types of potential hazards:
   (1) criticality
   (2) chemical hazards
   (3) radiological hazards
   (4) fires and explosions

8   Word Usage

The following pairs of words are frequently misused or confused.

accept/except

To accept is to receive, agree to, or consider proper, right, or true. As a verb, to except means leave out or exclude; as a preposition, it means excluding.

- You must accept the responsibility that goes with the appointment.
- We agreed on everything except the schedule.

adapt/adopt

To adapt is to adjust or make suitable. To adopt is to accept or make one's own.

- He will adapt to the motion of the sea in a few days.
- I will adopt your agenda for the meeting.

advice/advise

Advice is a noun and advise is a verb.

- My advice is to sign the contract immediately.
- I advise you to sign the contract immediately.

affect/effect

Effect is often misused for the verb affect, which means to have an effect on or to influence. To effect is to bring about.

- The decisions of the public utility commission affect all State utilities.
- These policy changes had a good effect on staff morale.

alternate/alternative

An alternate is a substitute, an alternative is a choice between two or more possibilities. As an adjective, alternate means "by turns" or "every other."

- He appointed Arthur as the alternate.
- The inspector had no alternative to shutting down the plant.

among/between

Use between for two persons or things; use among for three or more.

- This discussion is between you and me.
- The three technicians discussed the test results among themselves.

assure/ensure

To assure is to set a person's mind to rest and is used to address people. To ensure is to make certain.

- I assure you that the documents will arrive on schedule!
- Adhering to this maintenance schedule will ensure proper operation of the system.

beside/besides

Beside is a preposition and besides is an adverb.

- Place the telephone beside the bookcase.
- Besides, he knew when the meeting was to convene.

compose/comprise

Comprise means to include or contain. Do not use the construction "comprised of"; use "composed of." The whole comprises the parts.

- The committee comprises 12 members and three subcommittees.
- Each panel is composed of six experts in health physics.

due to/because of

Because of is an adverbial prepositional phrase; due to is an adjectival prepositional phrase. Only use due to to modify a noun or after a form of the verb to be.

- The delay was due to an automobile accident.
- Because of the delay, the schedule was changed.

farther/further

Use farther for geographical distance and further in more abstract senses.

- He moved 2 kilometers (km) farther down the road. His first office was only 1 km away.
- She must study further to achieve mastery of the subject.

fewer/less

Use fewer for countable quantities; use less for qualities or quantities that cannot be counted individually. Also use less for time and money.

- Fewer people attend the meeting each year.
- Less activity than predicted was visible around Jupiter's moon.

practical/practicable

According to Webster's II New College Dictionary, "*Practicable* refers to something that can be put into effect. *Practical* refers to something that is also sensible and worthwhile."

- While it may be practicable to give every employee a car and cell phone, it is neither cost efficient nor practical.
- After a review of the plans, the staff believes the expedited schedule for construction is ambitious but still practicable.

principal/principle

As an adjective, principal means "first" (in importance or degree). A principle is a basic truth, rule, or standard.

- The principal speaker was the Director of NRR.
- We follow the NRC Principles of Good Regulation.

prior to/before

Prior is an adjective. Do not use "prior to" as a preposition. Use "before."

- The agreement signed today replaces the prior agreement.
- The report arrived before the letter.

shall/will/must/may/may not/should

In regulations, "shall" and "will" indicate a requirement, "may" an option, and "may not" a prohibition. "Shall" and "will" are used for persons or organizations; "must" for inanimate objects. In guidance such as regulatory guides, "should" is often used for recommendations.

- The licensee shall check the operation of reusable collection systems each month.
- The program plan must describe the licensee's procedures.
- The licensee may submit the report by electronic mail.
- A licensee may not administer that dose to humans.
- The licensee should carry out these drills before the actual test.

that/which

Do not use the relative pronoun "that" to begin a nonrestrictive clause (a clause that can be omitted without changing the meaning of the sentence). Many writers reserve "which" (preceded by a comma) for nonrestrictive clauses.

- Training on a simulator ensures a degree of competence that cannot be learned otherwise.
- She walked to the warehouse, which is a mile away, to pick up the reports.

9   Footnotes and Credit Lines

Use footnotes to supply explanatory material or information that would interrupt the flow of ideas in the text and when too few footnotes exist to warrant listing them as an appendix to the text. Use footnotes to a table or figure to clarify or elaborate on the data in the table or figure.

1. **Place a footnote** on the same page or under the same column as its reference.

   - The Advisory Committee on the Medical Use of Isotopes (ACMUI) gives expert opinion on the medical use of radioisotopes to staff at the NRC.[1]

   ---
   [1]   The Atomic Energy Act of 1954, as amended in 1974, limits the NRC's regulation of radioactive materials to reactor-produced isotopes.

2. **Place a footnote reference** (i.e., a superscript number, symbol, or letter) applicable only to information in parentheses or brackets **inside the closing parenthesis or bracket.**

   - The position closed April 4, 2009, without a suitable applicant chosen (Six engineers applied.*)

3. Place a footnote reference outside all punctuation except the dash.

   - The trip from the site to the airport **—about 60 miles—was slow and tiring.

4. **Separate a footnote reference from the word preceding it** by a thin space unless it is preceded by a punctuation mark (e.g., a period, comma, quotation mark).

   - The NRC* regulates this nuclear waste, and its container bears a classification label.

5. Separate footnote references occurring together by a comma and a thin space.

   - This topic was discussed in the previous three meetings. [1,2,3]

6. **Place footnotes to a table or figure**, which are handled independently from footnotes to text, immediately under the table or figure.

7. **Use a symbol or letter** rather than a number **for a footnote reference** if a number could be confused with the content of the table.

8. In a **table, place a footnote reference** to the right of a column of text or symbols and to the left of a column of figures.

Category	No. of associated reports
Personnel radiation exposures...............................	39
Lost, abandoned, and stolen material......................	*81
Leaking sources..	**15
. . .	
Other***..	67
Total	303

*   An NRER database record may be associated with more than one category of event.

** These numbers would be significantly higher if all lost or leaking static elimination devices had been reported (see Section 2.1.1.3).

*** *Other* includes categories such as medical, transportation, and miscellaneous.

9. Place a **credit or source line for a table or figure**, which identifies where the data were obtained, directly under and flush left with the last footnote. For a figure or table without footnotes, place the source line directly under and flush left with the title.

- Figure 4.3. Chernobyl data evaluation of power vs. time during core destruction phase (Sheron, 1986)

  Source: Soviet analysis provided in Figure 4 of U.S.S.R, 1986

10   References

1 Purpose

A reference gives credit to an author for the information used in a document and directs the reader to the source of the information. NRC authors must provide sufficient information in each reference citation to enable the reader to obtain the referenced document from a location accessible to the public.

2 Availability

Include sources that are publicly available in a list of references. Do not include, for example, private communications, predecisional documents, technical notes, minutes of a meeting, or other sources that are not available in the NRC's Public Document Room.

Avoid using a classified or proprietary document as a reference unless it is the only source of information cited. If you must use such a document as a reference, state the following, as appropriate, after its citation in the list of references:

Classified report. Not publicly available.
or
Proprietary information. Not publicly available.

The author of a document is responsible for ensuring that each reference is accurate and that each document referenced is publicly available, unless otherwise indicated.

The NRC publishes an availability notice on the inside front cover of its reports (see Section 1 of the appendix to this guide). This notice directs the reader to sources for obtaining publicly available NRC documents and most codes and standards referenced in NRC reports. Therefore, the NRC does not include NRC documents as references but does include references to codes and standards as a separate category in a reference list for these reports.

The availability statements presented in Section 1 of the appendix to this guide cover appropriate sources for most references that would be included in NRC documents other than NUREG-series reports. Often these statements would be used as a footnote to a reference that is fully identified in the text.

3 Placement

- For most documents, place the list of references in a separate section immediately after the text.
- If there are too few references to warrant a list, include the references as footnotes or place them in parentheses directly in the text.

4 Legal Citations

Statutory material and court decisions are usually cited in the text or footnotes to a document rather than in a list of references. (However, see the examples of legal citations included in the sample lists of references in Sections 2 and 3 of the appendix to this guide.) For detailed guidance on citing legal materials, consult *The Bluebook: A Uniform System of Citation*, which is available from the NRC Law Library and most public libraries.

5 Listing and Identifying References

- Most references include the following information. Order the information appropriate to each reference as follows, and separate the components by commas:
  - **Author**—List the authoring individual, agency, corporation, or association.
  - **Title**—Italicize the title of a book and enclose the title of a journal article or the chapter of a book in quotation marks. Place the title of a journal article before the italicized name of a journal. Enclose the title of a technical report, a regulatory guide, and an industry code or standard in quotation marks.
  - **Publisher and location**
  - **Volume**
  - **Page number**
  - **Date**

5.1 Listing by Category of Document

For a NUREG–series report that has an availability notice on the inside front cover, identify the references in the text parenthetically or directly as part of a sentence, and list the references by category, alphabetically, and, as appropriate, sequentially by number. Identify NRC documents in the text, but do not include them in the list of references. Do not include the publisher and location for documents covered by the availability notice. (See Section 2 of the appendix to this guide for examples of how to list and identify references in these reports.)

Listing references

- Categorize the references by name of author or type of document (e.g., correspondence, codes and standards), but do not include NRC documents in the list of references.
- Under the category for industry codes and standards, alphabetize the subcategories (e.g., ANS, IEEE) and then list the individual references sequentially by code or standard number. Place this category last in the list of references.
- For the first document listed in each category, give all the components of information for a reference except the publisher and location for documents covered by the availability notice, and for each subsequent document, give only its alphanumeric designator, title, and date, in that order.
- List the references alphabetically or, as appropriate, sequentially by number.

- List a single author or the first author, if more than one, by last name, first initial, middle initial, if available. List subsequent authors by first initial, middle initial, last name. For more than three authors, follow the first author by et al.

Identifying references in the text

- Identify a reference in the text so that it is easy to find—in parentheses by an alphanumeric designator or some other descriptive information, or directly in a sentence:
    - (NUREG–0800)
    - (NRC Bulletin 89–11)
    - (ADAMS Accession No. ML080180332) *but*—define Agencywide Documents Access and Management System (ADAMS) for the first reference.
    - (5 U.S.C. 553, 555)
        - ❖ Good cause exists under 5 U.S.C. 553(d) to dispense with the usual 30-day delay in the effective date of the final rule.
        - ❖ This final rule does not contain information collection requirements and, therefore, is not subject to the Paperwork Reduction Act of 1995 (44 U.S.C. 3501 et seq.).
    - (Smith, 1987)
    - The evaluation (Smith and Jones, 1987) showed that....
    - The report by Smith and Jones (1987)....
    - Smith and Jones (1987) reported....
- Identify a reference in the text in the same way in which it appears in the list of references. For example, (Smith, 1987) should appear in the list of references under Smith's name rather than under the number of Smith's report.
- Identify multiple documents written by one author first by date, and then alphabetically by lowercase letter: (Smith, 1987), (Smith, 1988a), (Smith, 1988b).
- Cite a particular page, chapter, figure, table, or equation in the text rather than in the list of references, if possible.

5.2 Listing Sequentially by Number

For a document in which the references are numbered, give all the components of information for each reference (see Item 5 of this section), and list the references in the same order that they appear in the text. Identify the references in the text parenthetically by number. Note that the sample numbered references in Section 3 of the appendix to this guide include NRC documents and, therefore, are not representative of a references list that would be included in an NRC NUREG-series report.

The NRC discourages the use of numbered references in long documents. NRC documents often must be prepared and published in short turnaround time to coincide with licensing actions and are often changed as a result of multiple regulatory reviews. These conditions can result in last-minute renumbering of references. This guidance does not apply to correspondence.

Listing references

- List the references in the same order that the information referenced appears in the text (i.e., sequentially by number).
- Order the components of information appropriate to the references.
- List all authors by first initial, middle initial if available, and last name.

Identifying references in the text

- Avoid numbered references for long documents. A long document may have hundreds of references and any change in the text could require extensive renumbering of the references. In a long document, use a more stable method of referencing such as (Author, Year). For short documents, numbered references may be acceptable. When writing for an outside publisher or conference that uses numbered references, identify a reference parenthetically in the text, using Arabic numerals. You may also abbreviate the word "reference": (Ref. 1) or (Refs. 3–8). This guidance does not pertain to correspondence. See Management Directive 3.57, "Correspondence Management," for guidance on identifying references in correspondence.

6 Cross-References

- To cross-reference, direct the reader to another section of the same document by using an in-text citation such as (see Appendix B), (see Table 3), or (see Section 9.2 of this guide). However, avoid vague, nonspecific cross-references such as (see passages on high-level waste in this document).
- Cite the cross-reference in the same form as the entry cross-referenced. For example, if "Section 3" is cross-referenced, write (see Section 3), *not*— (see Section III).

11 *Federal Register* Documents

The Office of the Federal Register (OFR) requires that rulemakings, petitions for rulemakings, and general notices submitted for publication in the *Federal Register* comply with its format requirements. The following basic style guidance is consistent with these OFR requirements. For additional specific procedural guidance, consult NUREG/BR–0053, Revision 6, "United States Nuclear Regulatory Commission Regulations Handbook."

1. **Capitalize** only the **first word of a section heading**, unless the components of the section heading are separated by a colon. Place a period at the end of the section heading.

   - § 4.62 Right to counsel.
   - § 53.13 Contents of a request: Alternatives.

2. **Capitalize the first word of each item in a list** that follows a colon or a dash (use a colon following a complete statement and a dash following a phrase). **Use a semicolon after each item in a list** and a conjunction after the next-to-last item in the list unless each item in the list is a complete sentence.

   - § 53.12 Contents of a request.
     A request for a Commission determination under this part must include the following general information:
       - (a) Name and address...is sought;
       - (b) The civilian nuclear...is being made; and
       - (c) Explanation of why...is needed.

   - § 55.57 Renewal of a license.
       - (a) applicant for renewal of a license
     shall—
       - (1) Complete and sign Form 38.
       - (2) File an original form and two copies as
     specified in § 55.5(b).

3. **Capitalize the first word of terms to be defined** in a Definitions section. Enclose the terms to be defined in quotation marks, and list the terms, without designation, in alphabetical order.

   - "Buffer zone" means....
   - "Active maintenance" means....
   - "Commission" means....

4. Capitalize the first word of a table or figure and all proper nouns in the title of a table or figure. Also capitalize the first word of a column heading.

- Figure 1. Criteria for free-drop test
- Table 3. Estimates of truck accidents in Virginia
- <u>Package weight</u>

    <u>Fragility</u>

    <u>Free-drop distance</u>

5. Use the following format when referring to a **title or portion of a title** of the *Code of Federal Regulations* (see also Section 5, Item 3, of this guide).

- Title—Title 10 of the *Code of Federal Regulations*
- Chapter—10 CFR Chapter 1
- Part—10 CFR Part 50
- Section—10 CFR 50.16
- Appendix—10 CFR Part 50, Appendix R, *or* Appendix R to 10 CFR Part 50

6. **Use the following conventions** for documents to be published in the *Federal Register:*

FR	*Federal Register*
CFR	*Code of Federal Regulations*
U.S.C.	*United States Code*
Pub. L.	Public Law
Stat.	*U.S. Statutes at Large*
E.O.	Executive Order
Proc.	Proclamation
sec.	Section of a statute.
§ 2.8	Section of a regulation (except use the word rather than the symbol to begin a sentence). Place a space between the symbol and figure.
DC	District of Columbia, unless part of a mailing address requiring a ZIP Code, then use DC.
NW.	Section within DC or other cities

7. **To indicate a requirement** in a rule, use *shall* with a person or organization and *must* with an inanimate subject. **To indicate a prohibition**, use *may not*. This usage of these words is **applicable to NRC management directives**.

   - The licensee shall record the data in a log.
   - The data must include the date and purpose of the visit and the visitor's name and affiliation.
   - The visitor may not enter any vital area.

8. When submitting **documents to be published in the *Federal Register*,** the Office of the Federal Register prints these words in their own format, regardless of how they are submitted. When referring to the *Federal Register* in **other documents**, it is appropriate to **italicize *Federal Register*** because it is in the title of a publication.

9. When citing a part of the *Federal Register* in text, use the following format: 54 FR 33168; August 11, 1989.

   - The basis for the NRC's occupational chemical toxicity limits for uranium is given in an amendment to 10 CFR Part 20 (39 FR 13671; April 16, 1974) and is based on the threshold limit value.
   - On August 28, 2007 (72 FR 49352), the U.S. Nuclear Regulatory Commission (NRC) published a final rule revising the provision.

12   Technology and Writing

Although writers may still brainstorm with pen and paper, all agency documents are now completed using a computer word processing program. Word processing programs are constantly improving, offering authors powerful new tools for writing, editing, and collaborative work. In this section, we present tips for getting the most out of writing and editing on a computer as well as ways to ensure that your final document is free of editing marks, comments, and other private information.

We are offering general guidance for word processing. Because technology changes, we are not offering step-by-step guidance. Please use the Help menu on your word processing toolbar to search for more detailed instructions.

1 Start Writing

- Before you begin work on a document, make sure you have the **proper template**. Certain regulatory documents have specific styles that dictate different font sizes, bullet styles, and margins. Also, make sure that your computer word processing program is set **to follow basic NRC font type, font size, and margins**.

- As you write, you may want to highlight areas for review by using the **Highlight** function. If you would like to make notes about the area, use the **Comment** function to flag it with a comment bubble. You can make notes in the comment bubble. Right click on a comment to delete comment bubbles individually or use the **Delete All Comments in Document** function.

- To ensure you do not lift outside styles and formatting, use **Paste Special** and then choose **Unformatted Text** (instead of **Paste**). If you need to lift sections of text from another document for direct quotation in your own document, select the desired text in the source document, then **Copy** (Ctrl+C). Go back to your document, place your cursor where you want the text and use **Paste Special** (*not* regular **Paste** or Ctrl+V), and choose **Unformatted Text** to put down the text without introducing foreign styles.

- When writing a list, **use automated formatting to generate automatic bullets**, numbers, letters, Roman numerals, or outlines.

- **Use nonbreaking spaces** (Ctrl+Shift+Spacebar) to control line breaks in dates and between a word and related number. For example if you cite "Table 3.1," use a nonbreaking space to ensure that "Table" and "3.1" do not become separated on two lines.

- **Use nonbreaking hyphens** (Ctrl+Shift+Hyphen) within all document numbers.

2 Edit Efficiently

- When editing a document, turning on **Track Changes** allows you and the author to see each change. Insertions and deletions are clearly shown through redlines, strikeouts, underlines, and comment bubbles.

- Authors can right click to accept or reject changes one by one. Authors can also accept or reject all changes in the document at once.

- Use the **Compare and Merge Documents** function to show all differences between the edited document and the original document.

3 Clean Your Final Document

- Do not let editing changes hide in your final document. The **Display for Review** function will appear to accept all editing changes if you choose **Final** (instead of **Final Showing Markup**). However, anyone will be able to see the original document or the document with editing marks by switching the display back to **Original** or **Final Showing Markup**.

- To clear editing, follow the earlier directions for accepting or rejecting all changes in the document. Do not choose **Final** in the **Display for Review** function and then consider the document "cleaned."

- Use the **Remove Hidden Data** function from the **File** menu. After data are removed, resave the cleaned document as your final copy.

4 Check the Grade Level of Your Writing

- If you are writing a document for a nontechnical audience (such as a document for public comment), **aim for grade 9 to 12 writing**. Microsoft Word can automatically check the reading level of your document with a formula that uses the average number of syllables per word and the average number of words per sentence.

- Go to **Tools** and click **Options**. Then choose the **Spelling and Grammar** tab. Check the **Check grammar with spelling** box. Check the **Show readability statistics** box. Click **OK**.

- Run a normal spell check. At the end, you will see a measurement of your document's readability.

13 Writing in New Media

Much of the communication that once occurred over the phone or on paper is now happening through e-mail or the Web. The basic guidance for writing in new media is the same as for any good writing: write simply and use the principles of plain language. However, each medium also benefits from a few specific guidelines.

Writing E-Mails

Follow these six suggestions to help avoid misunderstandings over e-mail.

1. **Cover one topic in each e-mail**. Instead of saving up all questions and concerns for one long e-mail, keep e-mails short and address only one, or at most two, topics in an e-mail.

2. **Control the recipients**. CC only those with a legitimate need to know.

3. **Use an accurate subject line**. Keep your subject line descriptive and short.

4. **Keep the tone professional**. In person or on the phone, you can modulate your message with a cheerful voice or a smile. E-mail offers only words. What you may think is direct, the reader might find blunt or even rude. Words like "please" and "thank you" help control tone.

5. **Do not use all capital letters**. Many readers interpret this as being yelled at.

6. **Be careful what you write**. Although you may think you have deleted an e-mail, an e-mail never disappears. It will continue to exist and may be searched for and found.

Writing for the Web

The Web is an essential medium for communicating with the NRC's internal and external stakeholders. Use the following general tips for writing for the Web. For more detailed information on developing Web content for the NRC's external Web site, see http://www.internal.nrc.gov/web-standards.

1. **Answer the user question, "Why should I read this page?"** Begin your Web page or site with a short introduction that states what topics the site covers and why.

2. **Write at a ninth-grade reading level.** The NRC public Web site must serve the needs of a diverse audience. Use plain, conversational language. To determine the grade level of your writing, see Section 12, "Technology and Writing," in this guide.

3. **Organize information**.

- Do not overload users with information.
- Start new pages or topics with a brief introduction to provide context. Remember that users may not proceed through your Web pages from beginning to end but may skim only certain pages or may arrive at any page through a search.
- Structure each page for quick scanning using clear headings. Use short sentences (no more than 20 words) and small paragraphs (no more than 6 sentences). About 80 percent of users scan any new page. Only 16 percent read each word. Readers can easily miss information when a page contains dense text.
- Avoid large blocks of text.

4. **Provide links**. Guide users to supportive information such as a glossary, a frequently asked questions (FAQ) section, outside sources, and additional information.

5. **Communicate effectively**.

- Compose sentences in active rather than passive voice.
- Write instructions in positive rather than negative statements. For example, "Write in short sentences," is better than, "Do not write in overly long sentences."
- Use the principles of plain language. See Section 7, "Plain Language," in this guide.
- Use acronyms or abbreviations sparingly. Define an acronym on first mention, but remember that users may easily miss the definition if they scroll past it or enter the page below the definition. Show complete words rather than abbreviations whenever possible. Only use abbreviations when they are significantly shorter than the original term and easily understood by the typical user.
- Do not use words or phrases that a typical user may not understand. For example, some users may have trouble understanding "radon screening." Changing the text to "testing for radon" will substantially improve users' understanding. To go a step further, "testing for radon, a colorless, odorless, radioactive gas present in many homes," is even easier to understand. Adding a glossary may be helpful but is not an excuse to pepper the text with complex terms.

6. **Be current and credible**.

- Provide articles containing citations and references.
- Show the credentials of an author whose work you have quoted or summarized.
- Ensure the site is as up-to-date as possible. Put the last revised date on the site.

14   Helpful Web Reference Tools

Glossary of Nuclear Terms

http://www.nrc.gov/reading-rm/basic-ref/glossary.html

NRC Collection of Abbreviations

http://www.nrc.gov/reading-rm/doc-collections/nuregs/staff/sr0544/r4/

U.S. Government Printing Office Style Manual

http://www.gpoaccess.gov/stylemanual

NRC Plain Language Writing Techniques

http://www.internal.nrc.gov/ADM/DAS/cag/notices/notdocs/writingtech.html

Correspondence Management (Management Directive (MD) 3.57)

(Scroll down to MD 3.57 for guidance on writing letters for the NRC)

http://www.nrc.gov/reading-rm/doc-collections/management-directives/volumes/vol-3.html

NRC Regulations Handbook

(Guidance on rulemaking, including writing techniques and sample documents)

http://www.internal.nrc.gov/ADM/DAS/cag/RM01/handbook/index.html

Office of the Federal Register Document Drafting Handbook

http://www.archives.gov/federal-register/write/handbook/

The NRC Rulemaker: Web Assistance for Rulemaking Staff

(The Drafting Tools section includes sample documents, templates, and writing guidance.)

http://www.internal.nrc.gov/ADM/DAS/cag/RM01/draft.html

Information Digest

http://www.nrc.gov/reading-rm/doc-collections/nuregs/staff/sr1350/

*URLs were functional at time of printing. Please note that URLs may change.

Appendix
Sample NUREG

Contents

1 Availability

Availability notices inform the public on accessing NRC materials.

1.1 Notice from a NUREG-Series Report

This is a notice of availability for a NUREG-series report.

AVAILABILITY OF REFERENCE MATERIALS
IN NRC PUBLICATIONS

NRC Reference Material

As of November 1999, you may electronically access NUREG-series publications and other NRC records at NRC's Public Electronic Reading Room at http://www.nrc.gov/reading-rm.html.
Publicly released records include, to name a few, NUREG-series publications; *Federal Register* notices; applicant, licensee, and vendor documents and correspondence; NRC correspondence and internal memoranda; bulletins and information notices; inspection and investigative reports; licensee event reports; and Commission papers and their attachments.

NRC publications in the NUREG series, NRC regulations, and *Title 10, Energy*, in the Code of *Federal Regulations* may also be purchased from one of these two sources.
1. The Superintendent of Documents
   U.S. Government Printing Office
   Mail Stop SSOP
   Washington, DC 20402–0001
   Internet: bookstore.gpo.gov
   Telephone: 202-512-1800
   Fax: 202-512-2250
2. The National Technical Information Service
   Springfield, VA 22161–0002
   www.ntis.gov
   1–800–553–6847 or, locally, 703–605–6000

A single copy of each NRC draft report for comment is available free, to the extent of supply, upon written request as follows:
Address: U.S. Nuclear Regulatory Commission
           Office of Administration
           Mail, Distribution and Messenger Team
           Washington, DC 20555-0001
E-mail: DISTRIBUTION@nrc.gov
Facsimile: 301–415–2289

Some publications in the NUREG series that are posted at NRC's Web site address
http://www.nrc.gov/reading-rm/doc-collections/nuregs
are updated periodically and may differ from the last printed version. Although references to material found on a Web site bear the date the material was accessed, the material available on the date cited may subsequently be removed from the site.

Non-NRC Reference Material

Documents available from public and special technical libraries include all open literature items, such as books, journal articles, and transactions, *Federal Register* notices, Federal and State legislation, and congressional reports. Such documents as theses, dissertations, foreign reports and translations, and non-NRC conference proceedings may be purchased from their sponsoring organization.

Copies of industry codes and standards used in a substantive manner in the NRC regulatory process are maintained at—
           The NRC Technical Library
           Two White Flint North
           11545 Rockville Pike
           Rockville, MD 20852–2738

These standards are available in the library for reference use by the public. Codes and standards are usually copyrighted and may be purchased from the originating organization or, if they are American National Standards, from—
           American National Standards Institute
           11 West 42nd Street
           New York, NY 10036–8002
           www.ansi.org
           212–642–4900

Legally binding regulatory requirements are stated only in laws; NRC regulations; licenses, including technical specifications; or orders, not in NUREG-series publications. The views expressed in contractor-prepared publications in this series are not necessarily those of the NRC.

The NUREG series comprises (1) technical and administrative reports and books prepared by the staff (NUREG–XXXX) or agency contractors (NUREG/CR–XXXX), (2) proceedings of conferences (NUREG/CP–XXXX), (3) reports resulting from international agreements (NUREG/IA–XXXX), (4) brochures (NUREG/BR–XXXX), and (5) compilations of legal decisions and orders of the Commission and Atomic and Safety Licensing Boards and of Directors' decisions under Section 2.206 of NRC's regulations (NUREG–0750).

1.2 Statements

- Available from most public libraries.
- Available from most technical libraries.
- Unclassified NRC correspondence and Commission papers with SECY designators are available for inspection or copying for a fee in the NRC Public Document Room, One White Flint North, 11555 Rockville Pike (first floor), Rockville, MD 20852.
- A free single copy of draft (*document*) is available to the extent of supply by writing to the Distribution and Services Unit, U.S. Nuclear Regulatory Commission, Washington, DC 20555. A copy is available for inspection or copying for a fee in the NRC Public Document Room, One White Flint North, 11555 Rockville Pike (first floor), Rockville, MD 20852.
- A copy of (*document*) is available for inspection or copying for a fee in the NRC Public Document Room, One White Flint North, 11555 Rockville Pike (first floor), Rockville, MD 20852. Copies may be purchased from the U.S. Government Printing Office, P.O. Box 37082, Washington, DC 20013-7082 or the National Technical Information Service, U.S. Department of Commerce, 5285 Port Royal Road, Springfield, VA 22161.

2 Categorical References

Please follow NRC style in your references. However, be aware that scholarly journals, publishing houses, and conferences may have their own reference styles. Above all, be consistent in your references.

2.1 Sample List

Books, articles, and other documents that represent an individual's work should be cited with the individual's name. Documents with no author, such as an unbylined newspaper article, should begin with the article title.

1.	Statute (See Section 10, "References," for citing in text.)	*Administrative Procedure Act*, § 6,5 U.S.C. § 555 (1982) 22 U.S.C. § 2567 (Supp. 1, 1983).
2.	Journal article	Alvarez, E., "Quantum Gravity: An Introduction to Some Recent Results," *Reviews of Modern Physics*, 61:561–604.
3.	Book, one author	Cameron, I.R., *Nuclear Fission Reactors,* Plenum Publishing Co., New York, NY, 1982.
4.	Book, more than three authors	Bonney, T.B., et al., *Industrial Noise Manual,* American Industrial Hygiene Association, Akron, OH, 1975.

5.	Conference proceedings article	Coutant, C.C., "Striped Bass and the Management of Cooling Lakes" (S.S. Lee and S. Sengupta, eds.), *Proceedings of the 3rd Conference on Waste-Heat Management and Utilization, 11–13 May 1981,* Hemisphere Publishing Co., Washington, DC, 1982.
6.	Newspaper article, news service	"GSU Gets Second Chance to Put River Bend Cost in Rates" (Associated Press), *State Times,* Baton Rouge, LA, p. B5, September 21, 1989.
7.	Correspondence	James, Dale E., Entergy Arkansas Inc., letter to Ellis Merschoff, U.S. Nuclear Regulatory Commission, December 10, 1999, Agencywide Document Access and Management System (ADAMS) Accession No. ML003670601.
8.	Newspaper article, one author	Lippman, T.W., "Nuclear Exchange Brewing at NRC: Becquerel, Gray and Sievert May Obliterate Curie, Rad and Rem," *The Washington Post,* Washington, DC, p. A15, July 19, 1989.
9.	Book, compiled by editors	Sullivan, T.F.P., and M.L. Heavner, eds., *Energy Reference Handbook,* Government Institutes, Inc., Rockville, MD, 1981.
10.	Technical report, corporate author	Teknekron Research, Inc., "Utility Management and Technical Resources," Executive Summary and Vols. 1–3, McLean, VA, May 1980.
11.	Court decision	*Three Mile Island Alert, Inc., v. NRC,* 771 F. 2nd 720, 740 (DC Cir., 1985).
12.	Bill	U.S. Congress, House Committees on Energy and Commerce and Interior and Insular Affairs, "State Nuclear Safety Participation Act of 1987," 100th Cong., 1st sess., H.R. 499, January 7, 1987.
13.	Book, two authors	Waltar, A.E., and A.B. Reynolds, *Fast Breeder Reactors,* Pergamon Press, New York, NY, 1981.
14.	Public Law, Statutes at Large	*West Valley Demonstration Project Act,* Pub. L. No. 96–368, 94 Stat. 1347 (1980).

Industry Codes and Standards

ANSI Standard

American National Standards Institute, ANSI N45.2.9–1974, "American National Standard Requirements for Collection, Storage and Maintenance of Quality Assurance Records for Nuclear Power Plants," New York, NY.

ANS Standards

American National Standards Institute/American Nuclear Society, ANSI/ANS 3.1–1978, "Selection and Training of Nuclear Power Plant Personnel," ANS, LaGrange Park, IL.

— — — ANSI/ANS 3.5–1981, "Nuclear Power Plant Simulators for Use in Operator Training," ANS, LaGrange Park, IL.

— — — ANSI/ANS 6.4–1977, "American National Standard Guidelines on the Nuclear Analysis and Design of Concrete Radiation Shielding for Nuclear Power Plants," ANS, LaGrange Park, IL.

ASME Standard

American Society of Mechanical Engineers, *Boiler and Pressure Vessel Code,* 1986 edition, Section III, Subsection NCA, "General Requirements for Division 1 and Division 2," 1988 addenda, New York, NY.

IEEE Standards

Institute of Electrical and Electronics Engineers, IEEE Standard 279–1971, "Criteria for Protection Systems for Nuclear Power Generating Stations," Piscataway, NJ.

— — — IEEE Standard 323–1974, "IEEE Standard for Qualifying Class 1E Equipment for Nuclear Power Generating Stations," Piscataway, NJ.

— — — IEEE Standard 382–1972, "IEEE Trial-Use Guide for Type Test of Class I Electrical Valve Operators for Nuclear Power Generation Stations," Piscataway, NJ.

2.2 Sample In-Text Identification

Cite sources in text by using parentheses and identifying information such as the author's last name and year of publication or regulation name. Generally, do not use numbered references for longer NRC documents. Conferences and scholarly journals may have their own standards. This guidance does not apply to correspondence.

- In parenthesis: He presented this theory in his book on nuclear fission reactors (Cameron, 1982).
- In a sentence: The effects of mechanical aging have been considered in the design process and the specific qualification test program, using the guidance of NUREG–0588, IEEE Standard 323–1974, IEEE Standard 382–1972, and various operability test procedures.
- He presented this theory in his book on nuclear fission reactors (Ref. 1).
- Many documents have been published that are related to the accident (Refs. 8 and 11).
- Reference 3 discusses the effects of noise on plant personnel.

3 Sample List

Below is a sample list of citations that should provide a template for most of your references. Books, articles, and other documents that represent an individual's work should be cited with the individual's name. Documents like NUREGs and Commission papers are considered the work of the agency and thus "U.S. Nuclear Regulatory Commission should begin those citations."

1.	Book, one author	Cameron, I.R., *Nuclear Fission Reactors,* Plenum Publishing Co., New York, NY,1982.
2.	Book, two authors	Waltar, A.E. and A.B. Reynolds, *Fast Breeder Reactors,* Pergamon Press, New York, NY,1981.
3.	Book, more than three authors	Bonney T.B., et al., *Industrial Noise Manual,* American Industrial Hygiene Association, Akron, OH, 1975.
4.	Book, compiled by editors	Sullivan, T.F.P. and M.L. Heavner, eds., *Energy Reference Handbook,* Government Institutes, Inc., Rockville, MD, 1981.
5.	Journal article	Alvarez, F., "Quantum Gravity: An Introduction to Some Recent Results," *Reviews of Modern Physics,* 61:561–604.
6.	Newspaper article, one author	Lippman, T.W., "Nuclear Exchange Brewing at NRC: Becquerel, Gray and Sievert May Obliterate Curie, Rad and Rem," *The Washington Post,* Washington, DC, p. A15, July 19, 1989.

7.	Newspaper article, news service	"GSU Gets Second Chance To Put River Bend Cost in Rates" (Associated Press), *State Times,* Baton Rouge, LA, p. B5, September 21, 1989.
8.	NUREG–series staff report	U.S. Nuclear Regulatory Commission, "Report to Congress on Abnormal Occurrences, Fiscal Year 2006," NUREG–0090, Vol. 29, April 2007, Agencywide Document Access and Management System (ADAMS) Accession No. ML071080195.
9.	NUREG–series contractor report	U.S. Nuclear Regulatory Commission, "Tornado Climatology of the Contiguous United States," NUREG/CR–4461, Rev. 2, February 2007, ADAMS Accession No. ML070810400.
10.	NUREG–series draft report	U.S. Nuclear Regulatory Commission, "Draft Environmental Impact Statement for an Early Site Permit (ESP) at the Vogtle Electric Generating Plant Site—Main Report" (Draft Report for Comment), NUREG–1872, September 20, 2007, ADAMS Accession No. ML072410045.
11.	Draft regulatory guide*	U.S. Nuclear Regulatory Commission, "Application and Testing of Safety-Related Diesel Generators in Nuclear Power Plants," Division 1, Proposed Revision 4 to Regulatory Guide 1.9, March 2007, ADAMS Accession No. ML062650307.
12.	Final regulatory guide*	U.S. Nuclear Regulatory Commission, "Reactor Coolant Pump Flywheel Integrity," Regulatory Guide 1.14, ADAMS Accession No. ML003739936.
13.	Commission Paper	U.S. Nuclear Regulatory Commission, "2007 Annual Report on Commission Adjudication," Commission Paper SECY–08–0009, January 17, 2008, ADAMS Accession No. ML080180332.
14.	Correspondence	James, Dale E., Entergy Arkansas Inc., letter to Ellis Merschoff, U.S. Nuclear Regulatory Commission, December 10, 1999, ADAMS Accession No. ML003670601.

*Generally, do not give dates and revisions for regulatory guides because they are periodically revised. However, give the alpha-numeric task designator for a draft guide or the number of a proposed revision to avoid confusion among multiple proposed revisions.

15.	Federal regulation	*U.S. Code of Federal Regulations*, "Domestic Licensing of Production and Utilization Facilities," Part 50, Chapter I, Title 10, "Energy."
16.	*Federal Register* notice (See Section 11, *"Federal Register* Documents," for in-text citation.)	U.S. Nuclear Regulatory Commission, "Rules of Practice for Domestic Licensing Proceedings & Procedural Changes in the Hearing Process (10 CFR Part 2)," *Federal Register,* Vol. 54, No. 154, August 11, 1989, pp. 33168–33182.
17.	Applicant and licensee document	Ebasco Services Incorporated, "Generic Issues Report—Evaluation and Resolution of Generic Technical Issues for Conduits and Conduit Supports," Rev. 2, Dockets 50–445/446, March 30, 1987.
18.	National standard	American Nuclear Society, "Selection and Training of Nuclear Power Plan Personnel," ANSI/ANS 3.1–1978, LaGrange Park, IL.
19.	National code	American Society of Mechanical Engineers, *Boiler and Pressure Vessel Code,* 1986 edition, Section III, Subsection NCA, "General Requirements for Division 1 and Division 2," 1988 addenda, New York, NY.
20.	Technical report	Teknekron Research, Inc., "Utility Management and Technical Resources," Executive Summary and Vols. 1–3, McLean, VA, May 1980.
21.	Conference proceedings article	C.C. Coutant, "Striped Bass and the Management of Cooling Lakes" (S.S. Lee and S. Sengupta, ed.), *Proceedings of the 3rd Conference on Waste-Heat Management and Utilization, 11–13 May 1981,* Hemisphere Publishing Co., Washington, DC, 1982.
22.	Bill	U.S. Congress, House Committees on Energy and Commerce and Interior and Insular Affairs, "State Nuclear Safety Participation Act of 1987," 100th Cong., 1st sess., H.R. 499, January 7, 1987.
23.	Public Law, Statutes at Large	West Valley Demonstration Project Act, Pub. L. No. 96-368, 94 Stat. 1347 (1980).
24.	Statute	*Administrative Procedure Act,* § 6, 5 U.S.C § 555 (1982) 22 U.S.C. § 2567 (Supp. 1, 1983).
25.	Court decision	*Three Mile Island Alert, Inc., v. NRC,* 771 F. 2nd 720, 740 (D.C. Cir., 1985).

64

Bibliography

NRC Guidance

U.S. Nuclear Regulatory Commission, "NRC Collection of Abbreviations," NUREG-0544, Rev. 4, July 1998.

— — — "Glossary of Terms," NUREG–0770, June 1981.

— — — "Preparing NUREG-Series Publications," NUREG–0650, Rev. 2, January 1999.

— — — "United States Nuclear Regulatory Commission Regulations Handbook," NUREG/BR–0053, Rev. 6, September 2005.

Other Guidance

American Chemical Society, "Handbook for Authors," Washington, DC, 1978.

American National Standards Institute, "Scientific and Technical Reports—Organization, Preparation, and Production," ANSI 239.18–1987, New York, NY.

Gibaldi, Joseph. *MLA Handbook for Writers of Research Papers*, Sixth Edition, Modern Language Association of America, New York, NY, May 2003.

National Archives and Records Administration, Office of the Federal Register, *Document Drafting Handbook,* U.S. Government Printing Office, Washington, DC, April 1986.

The Associated Press Stylebook, 42nd Edition, Basic Books, New York, NY, 2007.

The Chicago Manual of Style, 15th Edition, The University of Chicago Press, Chicago, IL and London, 2003.

U.S. Government Printing Office Style Manual, U.S. Government Printing Office, Washington, DC, 2008.

Words Into Type, Third Edition, Prentice-Hall, Inc., Englewood Cliffs, NJ, 1974.

Index

NRC FORM 335 (9-2004) NRCMD 3.7	U.S. NUCLEAR REGULATORY COMMISSION	1. REPORT NUMBER (Assigned by NRC, Add Vol., Supp., Rev., and Addendum Numbers, if any.)
	BIBLIOGRAPHIC DATA SHEET *(See instructions on the reverse)*	NUREG-1379, Rev. 2

2. TITLE AND SUBTITLE	3. DATE REPORT PUBLISHED	
NRC Editorial Style Guide	MONTH	YEAR
	May	2009
	4. FIN OR GRANT NUMBER	

5. AUTHOR(S)	6. TYPE OF REPORT
Caroline Hsu	
	7. PERIOD COVERED *(Inclusive Dates)*

8. PERFORMING ORGANIZATION - NAME AND ADDRESS *(If NRC, provide Division, Office or Region, U.S. Nuclear Regulatory Commission, and mailing address; if contractor, provide name and mailing address.)*

Division of Administrative Services
Office of Administration
U.S. Nuclear Regulatory Commission
Washington, DC 20555-0001

9. SPONSORING ORGANIZATION - NAME AND ADDRESS *(If NRC, type "Same as above"; if contractor, provide NRC Division, Office or Region, U.S. Nuclear Regulatory Commission, and mailing address.)*

Division of Administrative Services
Office of Administration
U.S. Nuclear Regulatory Commission
Washington, DC 20555-0001

10. SUPPLEMENTARY NOTES

11. ABSTRACT *(200 words or less)*

The "NRC Editorial Style Guide" provides writing and style guidance to all NRC staff members. It addresses the questions and issues most frequently fielded by NRC editors. This revision of the document emphasizes the importance of plain language and includes detailed guidance on using technology to write and edit more effectively. The goals of the "NRC Editorial Style Guide" are readability and consistency for all NRC publications.

12. KEY WORDS/DESCRIPTORS *(List words or phrases that will assist researchers in locating the report.)*	13. AVAILABILITY STATEMENT
NRC Editorial Style Guide Writing Plain Language Grammar Citations Word Usage Punctuation NRC publications NRC reports NRC editorial style	unlimited
	14. SECURITY CLASSIFICATION
	(This Page) unclassified
	(This Report) unclassified
	15. NUMBER OF PAGES
	16. PRICE

NRC FORM 335 (9-2004)

PRINTED ON RECYCLED PAPER

www.ingramcontent.com/pod-product-compliance
Lightning Source LLC
Chambersburg PA
CBHW081841170526
45167CB00007B/2865